Libraries and the Semantic Web

Synthesis Lectures on Emerging Trends in Librarianship

Editor

Hema Ramachandran, *California State University, Long Beach,* and **Joe Murphy,** *Director Library Futures, Innovative Interfaces and Technology Trend Analyst.*

This series, Emerging Trends in Librarianship (a sub-series of the Synthesis Lectures on Library Science and Librarianship), will focus on new and emerging trends in digital collections and new technologies as they relate to the practice of librarianship and library science. The series will be of interest not only to librarians and information professionals, but also to the research community in general. Topics include but are not limited to: eScience, institutional repositories, data curation, advances in discovery tools, taxonomy and thesauri construction, mobile technologies, and the newest topic of "near field communication."

Libraries and the Semantic Web
Keith P. DeWeese and Dan Segal
November 2014

Exploring New Roles for Librarians: The Research Informationist
Lisa Federer
April 2014

Near Field Communication: Recent Developments and Library Implications
Sheli McHugh, Kristen Yarmey
March 2014

Libraries and the Semantic Web
Keith P. DeWeese and Dan Segal

ISBN: 978-3-031-00910-5 print
ISBN: 978-3-031-02038-4 ebook

DOI 10.1007/978-3-031-02038-4

A Publication in the Springer series
SYNTHESIS LECTURES ON EMERGING TRENDS IN LIBRARIANSHIP #3

Series Editors: Hema Ramachandran, California State University, Long Beach and Joe Murphy, Librarian & Technology Trend Analyst

Series ISSN   2372-8833 Print   2372-8868 Electronic

Libraries and the Semantic Web

Keith P. DeWeese
Gannett Co., Inc.

Dan Segal
Marcinko Enterprises, Inc.

SYNTHESIS LECTURES ON EMERGING TRENDS IN LIBRARIANSHIP #3

ABSTRACT

This book covers the concept of the Semantic Web—what it is, the components that comprise it, including Linked Data, and the various ways that libraries are engaged in contributing to its development in making library resources and services ever more accessible to end-users.

KEYWORDS

bibliographic data, libraries, linked data, ontologies, RDF, semantic web, vocabularies

Dedications

Dedicated to Matthew Bloomfield-DeWeese.

Keith P. DeWeese

Dedicated to my family.

Dan Segal

Contents

Acknowledgments

Keith P. DeWeese

Thanks to my co-author, Dan, for his contributions to this book. I also wish to thank Hema Ramachandran, Kathleen Cottay, Ian Davis, and Joel Summerlin for their support and contributions to this book.

Dan Segal

Thank you to my co-author, Keith, for the opportunity to collaborate on this book. I also wish to thank my colleagues and mentors for their guidance and feedback, both on this book and on the work leading up to it, with special acknowledgement to: Daniela Barbosa, Christine Connors, Kathleen Cottay, Ian Davis, Beth Golden, Marti Heyman, Maureen Kelly, Sandra Kramer, Scott Lawrence, Randall Marcinko, Joel Summerlin, Harvey Wiener, and Trish Yancey.

Preface

This Synthesis Lecture is intended to provide an overview of the Semantic Web as it relates to libraries. It is for the librarian seeking to become familiar with and discuss the concept of the Semantic Web. It is also for the librarian engaged in applying procedures and processes, aligned to the concept of the Semantic Web and its standards, to their work whether in an effort to meet the needs of their end users, reduce administrative costs, or both.

Use-cases and examples primarily focus on bibliographic data for print resources and cover:

- standards and components that comprise the Semantic Web;

- an overview of Semantic Web projects that are underway at various libraries around the world;

- resources and tools, i.e., the technology of the Semantic Web; and

- Linked Data, a method of publishing structured data using Semantic Web practices so that the data can be interlinked and therefore become more useful (Wikipedia, 2014).

But why care about the Semantic Web and its life in libraries? To that, one might answer that, with the phenomenal and unceasing growth of content and data on the Web, all organizations must, in some capacity, process data by both human and machine reading while being engaged in improving the systems that access their data.

> *"When information about a library's collection is locked up behind a specific web site (such as an OPAC), it is often exceedingly difficult for services, such as search engines, that consume data. Information seekers need to be connected back to their*

local library resources from wherever they are on the web (OCLC, "Data Strategy and Linked Data," 2014).

CHAPTER 1

The Concept of a Semantic Web

With the birth and evolution of the World Wide Web, databases were allowed to proliferate beyond their closed and comparatively stable operational environments, whether that environment was a university or corporation, archive or library.

> "Databases today are made available, in some form, on the Web where users, application programs, and uses are open-ended and ever changing. In such a setting, the semantics of the data has to be made available along with the data. For human users, this is done through an appropriate choice of presentation format. For application programs, however, this semantics has to be provided in a formal and machine processable form. Hence the call for the Semantic Web" (Antoniou and Harmelen, 2004).

Distinctions between the Web and the Semantic Web can be summarized as follows.

The Web	The Semantic Web
The Web is an information space that contains information targeted at human consumption expressed in a wide range of natural languages	The Semantic Web is an information space in which information is expressed in a special machine-targeted language
The Web is a set of informally interlinked information.	The Semantic Web is a web of formally and semantically interlinked data
The Web is comprised of links	The Semantic Web is a "web of "data," which is the result of linking data (Bergsell, 2001)

It is important to note that linking data is not linking pages containing data, but actually linking one piece of information to another. With the Semantic Web, information exchange occurs as document and application layers are abstracted, i.e., as data structures are hidden. The Semantic Web, then, allows one to refer to specific pieces of information, even as they are updated, on pages or as part of applications (Eisenberg, 2011).

The World Wide Web Consortium states, "Linked Data lies at the heart of what [the] Semantic Web is all about: large scale integration of, and reasoning on, data on the Web (W3C, "Linked Data," 2014)." Accessing and integrating Linked Data at variously complex levels is achieved through use of standard technologies including the Dublin Core Metadata Element Set and the World Wide Web Consortium's Resource Description Framework (RDF), as we will discuss in later sections of this Lecture.

In 2006, Tim Berners-Lee published the first draft of a document titled, simply, "Linked Data" which, over time, has been edited, and in its latest draft, published in 2009, clarifies that the Semantic Web, unlike the "web of hypertext," is posited on four rules.

1. To be part of the discourse on the Semantic Web, a Web "thing" must be identified using Uniform Resource Identifiers or URIs. If a thing is not named using a URI, it's simply not a Semantic Web thing.

2. URIs must use HTTP so that the names can be looked up.

3. "Useful" information must be made available (and not, for example, hidden away in an archive) in standardized form, such as RDF, when a URI look-up occurs.

4. Links to URIs must be made so that additional, related things may be discovered (Berners-Lee et al., 2001).

Berners-Lee's four rules makes realizing the Semantic Web appear so easy; however, as practitioners, we know that, those rules are just a starting structure for understanding something that has been variously described as:

- an "extension of the World Wide Web that enables people to share content beyond the boundaries of applications and websites" (SemanticWeb.org contributors, 2012);

- "W3C's vision of the Web of Linked Data" (W3C, "Semantic Web, 2014);

- a means "to help computers 'read' and use the Web" (Wilson, 2006); and

- a web that, "describes the relationships between things" (Khan, 2014)."

Countless other descriptions exist, but what do any mean to librarians, what does the Semantic Web hold for librarians, and the current and future development of library management processes and user services as exemplified by practical application? To become involved in the Semantic Web *movement* and enter into its discourse, where does one begin? To find a beginning, we first examine the technical foundations of the Semantic Web, and then review selected Semantic Web initiatives undertaken by libraries around the world.

CHAPTER 2

Semantic Web Components

One approach to understanding the Semantic Web is to look at its components. The Semantic Web is built on standards for:

- naming information resources;

- classifying information resources;

- describing properties of resources;

- describing relationships of resources to other resources;

- representing knowledge in machine-readable format; and

- querying data sets.

The W3C describes this model as a "layer cake," as shown in Figure 2.1:

In our discussion, we will focus on the sections of the layer cake that are most relevant to librarians. The standards and concepts that librarians will most typically encounter when working with data in a Semantic Web include:

- Uniform Resource Indicators (URIs);

- Resource Description Framework (RDF);

- RDF Extensions: RDFS, OWL, and Specialized Vocabularies;

- SPARQL: Query Language for Semantic Web; and

- the Linked Data Cloud.

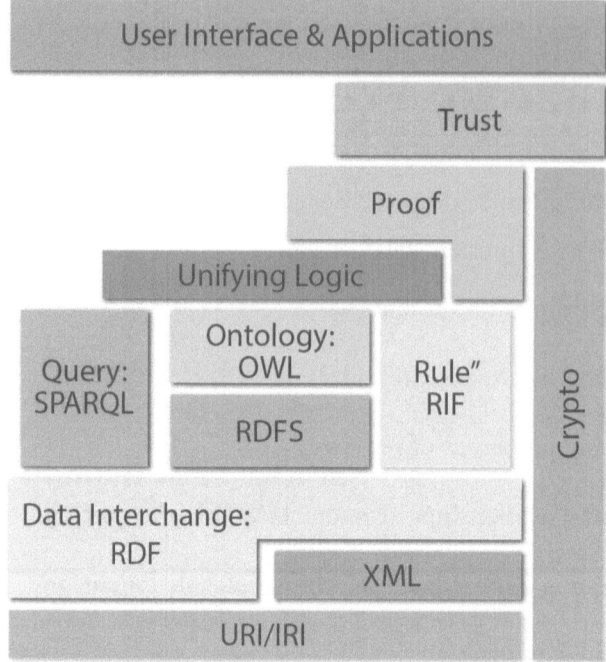

Figure 2.1: The layer cake model (based on: http://www.w3.org/2001/sw/).

2.1   UNIFORM RESOURCE INDICATORS (URIs)

The foundation of the Semantic Web is the Uniform Resource Indicator, or URI. A URI is formally defined as "a compact string of characters for identifying an abstract or physical resource (IETF, 2005)." In plain terms, a URI is simply a unique and Web-resolvable name for a data resource. This "data resource" can be any entity or any piece of useful information; for example, the name of an author, an organization, a geographic location, a published work, etc.

The generic URI syntax "consists of a hierarchical sequence of components referred to as the scheme, authority, path, query, and fragment (IETF, 2005)". Two example URIs and their component parts are shown below.

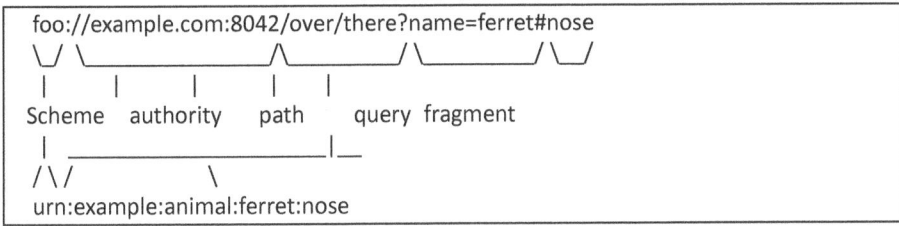

(based on: http://tools.ietf.org/html/std66#section-1.1.1)

The concept of naming data resources is not a Semantic Web innovation. Web users are already familiar with Uniform Resource Locators (URLs) and Uniform Resource Names (URNs), which are both types of URIs. For example, the URL http://www.worldcat.org/oclc/287628 uniquely refers to the OCLC WorldCat® entry for J.D. Salinger's novel *The Catcher in the Rye*. Similarly, the URN urn:isbn:9780316769532 uniquely identifies the published edition of *The Catcher in the Rye* by referencing its International Standard Book Number (ISBN): 978-0316769532.

A fundamental difference between the document Web and the Semantic Web is that in the former, unique ID's are assigned at the *document* level, while in the latter, unique ID's are assigned at the data *element* level. In other words, in the document Web, we can typically create links only between Web pages, files, and so on. However, in the Semantic Web, individual entities that occur within a document, such as person names, place names, etc., have unique identifiers, so that each entity can be referred to as a distinct unit. Hence, linking can occur at a much more granular level than just the document. The Semantic Web provides a framework where each attribute of an object is in itself a discrete piece of information that can be published, searched, or linked.

Author metadata is a good example. In the case of *The Catcher in the Rye*, the name of author J.D. Salinger occurs in several publicly available authority lists.

Authority	URI for J.D. Salinger
LC Name Authority File	http://id.loc.gov/authorities/names/n50016589
Virtual International Authority File (VIAF)	http://viaf.org/viaf/17092/
British National Bibliography	http://bnb.data.bl.uk/doc/person/SalingerJD(JeromeDavid)1919-2010
DBPedia	http://dbpedia.org/page/J._D._Salinger

Each of the URI's above represents the concept of "J.D. Salinger" in a different authority file. Thus, when creating a catalog record for *The Catcher in the Rye*, we are not constrained to filling in the Author metadata with a text string. We can also link our catalog record to an external source by referencing a URI from one or more of the published authorities. The mechanism for expressing the metadata values and creating the links is the *Resource Description Framework*.

2.2   RESOURCE DESCRIPTION FRAMEWORK (RDF)

Resource Description Framework, or RDF, is a W3C specification for describing Web resources. It is a standard model for data interchange (RDF Core Working Group). At the core of RDF is the notion of a "triple." A triple is a machine-readable statement that is comprised of:

- a Subject;

- a Predicate; and

- an Object.

Figure 2.2: An RDF "triple" (based on http://www.w3.org/TR/2004/REC-rdf-concepts-20040210/Graph-ex.gif).

In an RDF triple:

- the Subject is an information resource, which is typically represented by its URI;

- the Predicate, which is also represented as URI, denotes a relationship between the subject and is object; and

- the Object can be either a data value (for example, a text string, numeric value, date, etc.) or it can be a URI to another resource.

Hence, triples can describe absolute properties, as well as relationships between named resources. The collection of RDF triples is called a RDF graph.

To describe our sample work, we might wish to declare the following triples:

Subject	Predicate	Object
Resource	is type	http://schema.org/Book
Resource	has name	*The Catcher in the Rye*
Resource	has author	http://id.loc.gov/authorities/names/n50016589 http://viaf.org/viaf/17092/
Resource	is about	http://id.loc.gov/authorities/subjects/ sh2008112614
Resource	has publication year	1951
Resource	has ISBN	9780316769532

Graphically, our model would look like Figure 2.3.

In this representation, our Resource corresponds to the subject of each triple. Each arrow represents a predicate, and each oval target represents the object. Note that some of the objects have textual, numeric, or date values, while others have values that link to other data sets. For example, the name "J.D. Salinger" is expressed through a URI to the corresponding resources in the Library of Congress and VIAF authorities. Similarly, subject *Teenage boys—Fiction.* is expressed through a URI to the corresponding LC Subject Heading.

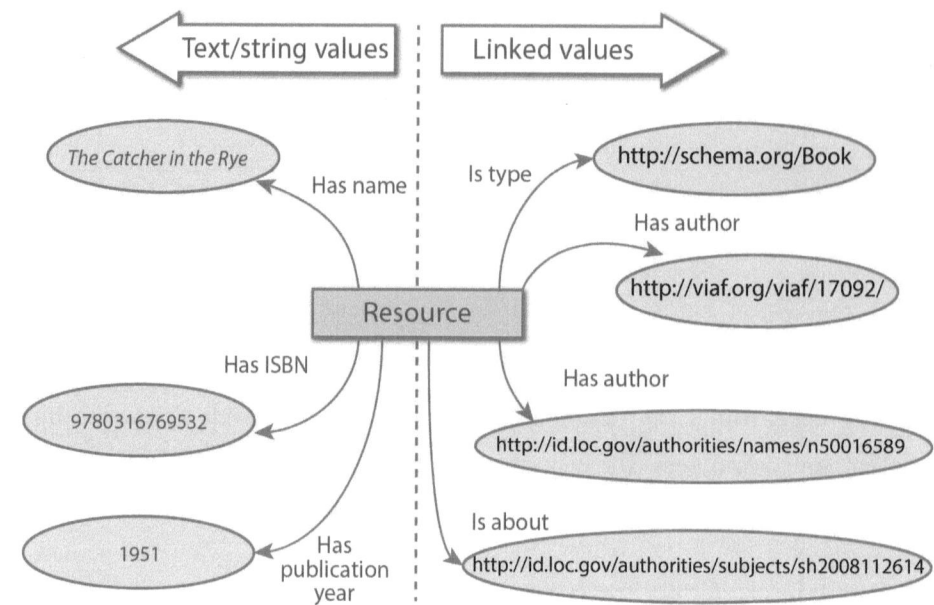

Figure 2.3: Our resource model.

To translate our model into machine-readable form, we need to assign a URI to our Resource. We will use http://worldcat.org/entity/work/id/375516175, which is the URI for the OCLC's semantic representation of *The Catcher in the Rye*.

In addition, we will reference some new URI's that are extensions of RDF, but which we will include here for illustration; for example:

- http://schema.org/Book, which indicates a value of "book" for the type of creative work;

- http://schema.org/name, a property that describes the name of an item;

- http://schema.org/author, a property that describes the author of a creative work;

- http://schema.org/about, a property that describes the subject matter of the content;

- http://schema.org/datePublished, a property that describes the date of first publication; and

- http://schema.org/isbn, a property that describes the ISBN of a book.

In machine-readable format, our graph appears as follows:

```
<rdf:RDF
 xmlns:rdf="http://www.w3.org/1999/02/22-rdf-syntax-ns#"
 xmlns:schema="http://schema.org/"
<rdf:Description rdf:about="http://worldcat.org/entity/work/id/375516175">
 <rdf:type rdf:resource="http://schema.org/Book"/>
 <schema:name>The Catcher in the Rye.</schema:name>
 <schema:author rdf:resource="http://viaf.org/viaf/17092"/>
 <schema:author rdf:resource="http://id.loc.gov/authorities/names/n50016589"/>
 <schema:about rdf:resource="http://id.loc.gov/authorities/subjects/sh2008112614"/>
 <schema:datePublished>1951</schema:datePublished>
 <schema:isbn>9780316769532</schema:isbn>
 </rdf:Description>
</rdf:RDF>
```

This graph is expressed, or serialized, in XML. However, RDF may be serialized in other formats as well, including N-Triple, JSON-LD, or Turtle, as shown below:

```
@prefix schema: <http://schema.org/> .
@prefix rdf:   <http://www.w3.org/1999/02/22-rdf-syntax-ns#> .

<http://worldcat.org/entity/work/id/375516175>
 a        schema:Book ;
 schema:name   "The Catcher in the Rye." ;
 schema:author <http://id.loc.gov/authorities/names/n50016589> , <http://viaf.org/
   viaf/17092> ;
 schema:about <http://id.loc.gov/authorities/subjects/sh2008112614> ;
 schema:datePublished   1951
 schema:isbn            9780316769532
```

Each of these formats and conventions is a standard syntax for data interchange on the Web. In our example, we invoked RDF, URI's, and the linking structure of the Web to populate the resource attributes for the type, author, and subject metadata—with values from remotely managed data sets. This approach hints at some of the potential benefits of linked data for libraries; for example, leveraging an "open, global pool of shared data that can be used and re-used to describe resources, with a limited amount of redundant effort (Library Linked Data Incubator Group, 2005)."

Before we begin to look at actual use cases, we need to emphasize that RDF is only the foundation. The RDF core enables us to define subjects, predicates, objects, and types. As we saw in our example, we need more than just the basic RDF expressions in order to describe the full range of properties that are associated with complex resources.

2.3   RDF EXTENSIONS: RDFs, OWL, AND SPECIALIZED VOCABULARIES

2.3.1   RDF SCHEMA (RDFs)

RDF Schema (RDFS) is an extension of RDF that "provides mechanisms for describing groups of related resources and the relationships between these resources" (W3C, 2014). More specifically, RDFS is "a vocabulary for describing properties and classes of RDF resources, with a semantics for generalization-hierarchies of such properties and classes" (W3C, 2004). RDFS builds on RDF by adding:

- subclasses; that is, the ability to place data resources into hierarchical groupings;

- sub-properties; that is, the ability to define hierarchical relationships for properties;

- annotations such as labels, comments, and definitions;

- domain specifications; which restrict the type of Subjects to which a particular property can be assigned; and

- range specifications, which restrict the type of Objects that can be assigned to a particular property.

These additions enable richer descriptions of information resources than we could achieve through subject, predicate, and object alone. Examples of some of the core RDFS components are shown below. Note that base of the URI—http://www.w3.org/2000/01/rdf-schema#—is represented by the prefix **rdfs**.

Label	URI	Description
Domain	rdfs:domain	A domain of the subject property
Member	rdfs:member	A member of the subject container
Range	rdfs:range	A range of the subject property
subClassOf	rdfs:subClassOf	The subject is a subclass of a class
subPropertyOf	rdfs:subPropertyOf	The subject is a sub-property of a property
comment	rdfs:comment	A description of the subject resource
isDefinedBy	rdfs:isDefinedBy	The definition of the subject resource
Label	rdfs:label	A human-readable name for the subject
seeAlso	rdfs:seeAlso	Further information about the subject resource

An example of where we see some of these components in use is BIBFRAME, the Library of Congress's semantic framework for bibliographic description and the planned replacement for MARC (Library of Congress, 2012). The BIBFRAME Vocabulary defines a general class of cataloging resources called Work. The Work class is then broken into specific subclasses such as Audio, Dataset, Moving Image, Text, etc.

Property	Label / Description		Subclass Of	
Audio	Audio / Resources expressed in an audible form, including music or other sounds.		Work	
Cartography	Cartography / Resource that show spatial information, including maps, atlases, globes,digital maps, and other cartographic items.		Work	
Dataset	Dataset / Data encoded in a defined structure. Includes numeric data, environmental data,etc., used by applications software to calculate averages, correlations, etc., or to produce models, etc., but not normally displayed in its raw form.		Work	
MixedMaterial	Mixed Material / Resource comprised of multiple types which are not driven by software. This may include materials in two or more forms that are related by virtue of their having been accumulated by or about a person or body, e.g. archival forms.		Work	
MovingImage	Moving Image / Images intended to be perceived as moving, including motion pictures (using liveaction and/or animation), film and video recordings of performances, events,etc.		Work	
Multimedia	Software or Multimedia / Electronic resource that is a computer program (i.e. digitally encoded instructions intended to be processed and performed by a computer) or which consist of multiple media types that are software driven. Examples include videogames and websites.		Work	
NotatedMovement	Notated Movement / Graphic, non-realized representations of movement intended to be perceived visually, e.g. dance.		Work	
NotatedMusic	Notated Music / Graphic, non-realized representations of musical works intended to be perceived visually.		Work	
StillImage	Still Image / Resource expressed through line, shape, shading, etc., intended to be perceived visually as a still image or images in two dimensions. Includes two-dimensional images and slides and transparencies.		Work	

Figure 2.4: A Bibframe example, based on: http://bibframe.org/vocab/Work.html.

In XML serialization, a class/subclass relationship appears as follows:

```
<rdfs:Class rdf:about="http://bibframe.org/vocab/Audio">
 <rdfs:label>Audio</rdfs:label>
 <rdfs:subClassOf rdf:resource="http://bibframe.org/vocab/Work"/>
 <rdfs:comment>Resources expressed in an audible form, including music or other
sounds.</rdfs:comment>
 </rdfs:Class>
```

Similarly, the BIBFRAME Vocabulary defines:

- properties that can be used for a given class;

- limits on the permitted values of those properties; and

- relationships to other properties.

These definitions can be used to ensure the quality of data records; for instance, by restricting which types of resources can be linked to one another.

For example, for the type of content object known as a Work, BIBFRAME defines a property called *absorbedBy*. This property indicates when a Work has been absorbed into, or superseded by, a different Work.

Property	Label / Description	Subproperty Of	Expected values
absorbed	Absorbed / Work that has been incorporated into another Work	precededBy	Work
absorbedBy	Absorbed by / Work that incorporates another work.	succeededBy	Work
absorbedInPart	Absorbed in part / Work that has been partially incorporated into another work.	precededBy	Work
absorbedInPartBy	Absorbed in part by / Work that incorporates part of the content of another work.	succeededBy	Work
accompaniedBy	Accompanied by / Resource that has an accompanying resource which adds to it	relatedTo	
accompanies	Accompanies / Resource that adds to or is issued with the described resource	relatedTo	

*Properties used with **Work** : Resource reflecting a conceptual essence of the cataloging resource.*

Resort a column by clicking the column header. Select a linked element below to learn more about that property or class.

Figure 2.5: Example 2 of a Bibframe, based on: http://bibframe.org/vocab/Work.html.

The BIBFRAME Vocabulary establishes that the *absorbedBy* property can be used only as a link between two Works. This type of constraint is called a *domain/range* constraint, where there is a defined scope for the link source, or domain (that is, the subject of the RDF triple), and the link target, or range (that is, the predicate of the triple). In XML, the property, its relationships, and its constraints are expressed as follows.

```
<rdf:Property rdf:about="http://bibframe.org/vocab/absorbedBy">
 <rdfs:domain rdf:resource="http://bibframe.org/vocab/Work"/>
 <rdfs:label>Absorbed by</rdfs:label>
 <rdfs:range rdf:resource="http://bibframe.org/vocab/Work"/>
 <rdfs:subPropertyOf rdf:resource="http://bibframe.org/vocab/succeededBy"/>
 <dcterms:modified>2014-04-10 (Updated subproperty)</dcterms:modified>
 <rdfs:comment>Work that incorporates another work.</rdfs:comment>
</rdf:Property>
```

2.3.2   WEB ONTOLOGY LANGUAGE (OWL)

At the next level of the Semantic Web stack, Web Ontology Language (OWL) extends RDF and RDFS. OWL "facilitates greater machine interpretability of Web content than that supported by XML, RDF, and RDF Schema (RDF-S) by providing additional vocabulary along with a formal semantics. OWL has three increasingly expressive sublanguages: OWL Lite, OWL DL, and OWL Full" (W3C, 2004). In addition, a next-generation Web Ontology Language, OWL 2, introduces new functionality and descriptive capabilities while remaining backwards-compatible (W3C, 2004).

OWL supports complex data models and granular resource description by introducing constructs such as:

- cardinality; that is, the minimum or maximum number of values that can be assigned to a given property;

- distinction between *object* properties—that is, a property that links a resource to one or more other resources, and *datatype* properties—that is, a property whose value is a literal, an integer, a date or time, a true/false value, etc.;

- advanced property types; for example, reflexive, symmetric, and transitive properties;

- disjoint classes;

- equivalent classes; and

- class unions and intersections.

The owl:sameAs property is an example of an OWL construct that is especially useful for Linked Data. This owl:sameAs expresses the relationship that two given individuals are equal:

Figure 2.6: The owl:sameAs property.

This type of triple can be used to express equivalency between resources in different data sets, and hence to create a link between the two, as well as a "hook" for integrating data from the different sources. The German National Library, for example, includes owl:sameAs references in its Integrated Authority File (GND), as shown below in the entry for Angela Merkel.

```
<rdf:Description rdf:about="http://d-nb.info/gnd/1051413044">
  <rdf:type rdf:resource="http://d-nb.info/standards/elementset/gnd#DifferentiatedPer
    son" />
  <gndo:gndIdentifier>1051413044</gndo:gndIdentifier>
  <gndo:variantNameForThePerson>Kasner, Angela Dorothea</gndo:variantNameForTh
    ePerson>
  <gndo:preferredNameForThePerson>Merkel, Angela</gndo:preferredNameForThePer
    son>
  <gndo:placeOfBirthAsLiteral>Hamburg</gndo:placeOfBirthAsLiteral>
  <owl:sameAs rdf:resource="http://www.filmportal.de/person/90be92b3a9044ef4bcf
    334cf997ae96f" />
  <gndo:gender rdf:resource="http://d-nb.info/standards/vocab/gnd/Gender#female" />
  <gndo:dateOfBirth rdf:datatype="http://www.w3.org/2001/XMLSchema#
    date">1954-07-17</gndo:dateOfBirth>
</rdf:Description>
```

Here, the link is to the record for Angela Merkel in the site filmportal.de. Other potential link targets include:

- other library authorities such as VIAF (http://viaf.org/viaf/12584821) or the British National Bibliography (http://bnb.data.bl.uk/doc/concept/person/lcsh/MerkelAngela1954-); and

- reference sources such as Dbpedia, a RDF-enabled version of Wikipedia content (http://dbpedia.org/page/Angela_Merkel), or Freebase, a "community-curated database of well-known people, places, and things" (http://www.freebase.com/m/0jl0g).

By establishing the semantic bridges between records for Angela Merkel in multiple sources, filmportal.de creates opportunities for mashing up content; for example, by enriching filmportal records with biographical information from DBpedia.

2.3.3   SPECIALIZED VOCABULARIES AND SCHEMA

In the Semantic Web, vocabularies "define the concepts and relationships used to describe and represent an area of concern" (W3C, "Vocabularies"). Vocabularies, also known as ontologies,[1] are built on RDF, RDFS, or OWL, with the addition of specialized class definitions, properties, and constraints that are needed to model a particular subject area or knowledge domain. Examples include the following.

- **Dublin Core Metadata Initiative (DCMI).** The Dublin Core Metadata Terms are available in RDF format from http://dublincore.org/ schemas/rdfs/. These include semantically expressed definitions of resource-descriptive metadata such as title, creator, date, publisher, etc.

- **schema.org.** A standard that is rapidly being adopted, schema.org "provides a collection of shared vocabularies Webmasters can use to mark up their pages in ways that can be understood by the major search engines: Google, Microsoft, Yandex, and Yahoo!"[2] Schema.org defines classes, properties, and constraints for describing resources such as creative works, events, organizations, people, places, and products. A common use case is to apply schema.org constructs to enriching Web publications with RDFa (Resource Description Framework in Attributes), a W3C Recommendation that adds structured metadata to HTML and XML. Note also that schema.org constructs are used extensively in the data model for OCLC's WorldCat Works (OCLC, "Data sets & services").

- **BIBFRAME.** As mentioned earlier, BIBRAME is a next-generation model for bibliographic description, as well as a planned replacement for MARC. Currently under development by the Library of Congress, BIBFRAME is based on a model of four core classes—Creative Work, Instance, Authority, and Annotation—with an extensive vocabulary of subclasses and properties (Library of Congress, 2012).

---

[1] Note that the W3C does not distinguish between the terms "vocabularies" and "ontologies." The trend is "to use the word "ontology" for more complex, and possibly quite formal collection of terms, whereas "vocabulary" is used when such strict formalism is not necessarily used or only in a very loose sense." (ibid.)

[2] Schema.org, n.d. Web. http://schema.org.

- **GeoNames.** Associated with the GeoNames database of over 10,000,000 geographic names, the GeoNames Ontology defines properties for describing geographic entities; for example, name, population, postal code, and neighboring entities.

- **Simple Knowledge Organization System (SKOS).** SKOS is a semantic framework that lends itself particularly well to semantic modeling of taxonomies and thesauri. It defines properties such as preferred label and alternative label, as well as relationships such as broader term, narrower term, related term, etc. The National Agriculture Library's Thesaurus and Glossary (http://agclass.nal.usda.gov/) is an example of a SKOS-based vocabulary that is available as Linked Open Data.

For the librarian, one of the advantages of vocabularies is that they can serve as off-the-shelf data models or data sets. Rather than "re-invent the wheel" by creating new, custom data models for common problems such as bibliographic description or knowledge representation, librarians can leverage existing vocabularies. Additionally, the use of standard, open vocabularies helps to ensure data quality and interoperability.

2.4   SPARQL: QUERY LANGUAGE FOR SEMANTIC WEB

Just as we use search engines such as Google to query documents, or query languages such as SQL to interrogate relational databases, we need a method to search and manipulate RDF data, so that we can extract useful information from RDF data sets. A practical use case might include, for example, generating lists from DBpedia, as we will see shortly.

The query language of the Semantic Web is called SPARQL (pronounced "sparkle"). While a discussion of SPARQL syntax is beyond the scope of this book, some key points to be aware of are:

- SPARQL searches RDF datasets (Feigenbaum and Prud'hommeaux, 2014);

- SPARQL queries are executed through a SPARQL endpoint, which serves as an interface into RDF data. The SPARQL endpoint can be either generic (that is, the endpoint interfaces to any Web-accessible RDF data) or specific (that is, the endpoint is preconfigured to work against particular datasets) (Feigenbaum and Prud'hommeaux, 2014). An example of a generic endpoint is OpenLink Virtuoso SPARQL Query Editor.

The following example illustrates a query of DBPedia for "all landlocked countries with a population greater than 15 million." [3]

```
SELECT ?country_name ?population
WHERE {
   ?country a type:LandlockedCountries ;
        rdfs:label ?country_name ;
        prop:populationEstimate ?population.

   FILTER(langMatches(lang(?country_name), "en")).
   FILTER (?population > 15000000) .
}
   ORDER BY DESC (?population)
```

The resulting output is a two-column table:

country_name	population
"Ethiopia"@en	93877025
"Afghanistan"@en	31108077
"Uzbekistan"@en	30183400
"Malawi"@en	16407000
"Burkina Faso"@en	15730977

2.5   THE LINKED DATA CLOUD

The W3C describes Linked Data as "the heart of what Semantic Web is all about: large scale integration of, and reasoning on, data on the Web (W3C)." Web data can consist of vocabularies and data models as described above, bibliographic records, reference works (for example, DBpedia, which makes Wikipedia infor-

---

[3] Adapted from Feigenbaum and Prud'hommeaux, 2014.

mation available in RDF format), data sets (for example, data.gov), semantically marked documents and publications, and more. The difference between a *Web* of data, and a mere *collection* of published data, is that in the former, data are available in a common format and can be queried, interrelated, and managed using standard technologies such as RDF, OWL, SKOS, SPARQL, etc. (W3C, "Linked Data").

The Linking Open Data (LOD) cloud diagram gives a visual sense of the size and interconnectedness of the Web of data. As of August 2011, 295 data sets had been identified, containing over 30 billion triples and 500 million outward links (LATC, 2014).

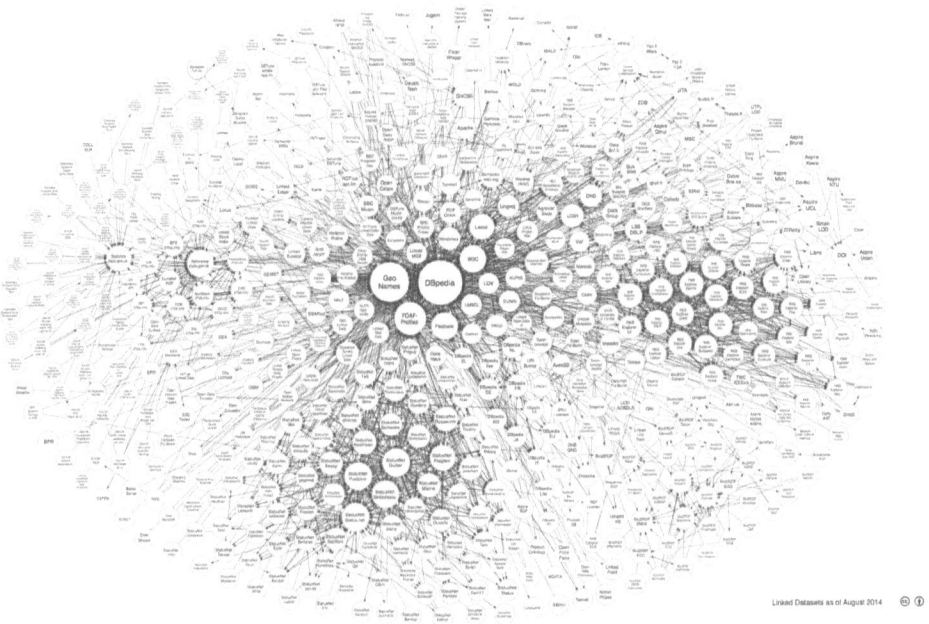

Figure 2.7: Linking Open Data cloud diagram 2014, by Max Schmachtenberg, Christian Bizer, Anja Jentzsch, and Richard Cyganiak, http://lod-cloud.net/.

Currently, a number of directories and search engines are available to help librarians locate linked, and linkable, data. Examples include:

- datahub. Allows users to search for data, register published datasets, create and manage groups of datasets, and get updates from datasets of interest;

- LinkedData.org. Presents lists of data sets;

- Linked Open Vocabularies (LOV). Lists and categorizes vocabularies. Supports search at the vocabulary or element level. Provides metrics about the use of vocabularies in the Semantic Web; and

- Sindice. Crawls the Web and indexes semantic sources. Offers search and querying across semantic data. Can be used to search at the vocabulary or element level.

Vocabularies member list:

Prefix	Namespace	Title
acco	http://purl.org/acco/ns#	Accomodation Ontology
acl	http://www.w3.org/ns/auth/acl#	Basic Access Control ontology
acm	http://acm.rkbexplorer.com/ontologies/acm#	ACM Classification Ontology
acrt	http://privatealpha.com/ontology/certification/1#	Agent Certification Ontology
ad	http://schemas.talis.com/2005/address/schema#	Address Schema
adms	http://www.w3.org/ns/adms#	Asset Description Metadata Schema
af	http://purl.org/ontology/af/	Audio Features Ontology
agls	http://www.agls.gov.au/agls/terms/	AGLS Metadata Terms
agrelon	http://d-nb.info/standards/elementset/agrelon.owl#	Agent Relationship Ontology
aliso	http://purl.org/vocab/aliso/schema#	Academic Institution Internal Structure Ontology
airs	https://raw.githubusercontent.com/airs-linked-data/lov/latest/src/airs_vocabulary.ttl#	Alliance of Information and Referral Services (AIRS) Vocabulary
akt	http://www.aktors.org/ontology/portal#	AKT Reference Ontology
akts	http://www.aktors.org/ontology/support#	AKT Support Ontology
algo	http://securitytoolbox.appspot.com/securityAlgorithms#	Algorithms Ontology
am	http://open-services.net/ns/asset#	OSLC Asset Management Vocabulary
ao	http://purl.org/ontology/ao/core#	Association Ontology
aos	http://rdf.muninn-project.org/ontologies/appearances#	Appearances Ontology Specification
api	http://purl.org/linked-data/api/vocab#	Linked Data API Vocabulary
apps4X	http://semweb.mmlab.be/ns/apps4X#	The vocabulary for Co-creation Events based on Open Data

Figure 2.8: LOV vocabularies member list (partial), http://lov.okfn.org/dataset/lov/index.html.

2.5.1 SUMMARY

In this chapter we examined some of the building blocks of the Semantic Web. The following chapters will look at some specific library initiatives in detail.

One standard by which any of these initiatives can be measured is Tim Berners-Lee's "Five Star" model for Linked Data, which awards "quality" stars as follows:

- One star: Data are available on the Web;

- Two stars: Data are in machine-readable format;

- Three stars: Data are published in a non-proprietary format;

- Four stars: Data representation complies with RDF standards; and

- Five stars: Data are linked using RDF (Berners-Lee, 2009).

For libraries, a benefit of producing "five-star" data is "to make the valuable information assets that (libraries) create and curate—resources such as bibliographic data, authorities, and concept schemes—more visible and re-usable outside of their original library context (Library Linked Data Incubator Group)". thereby extending the reach of the library and the size of its user community.

CHAPTER 3

Linked Data and Library Initiatives

Returning to Tim Berners-Lee's distinguishing the hypertext-based Web from the Semantic Web by way of the Four Rules, introduced earlier, there is a certain consolation in knowing that the rules are hard to break if for no other reason than "Breaking them does not destroy anything (Berners-Lee, 2009)." Not following them simply means that some opportunities might be missed in interconnecting data (Berners-Lee et al., 2001). This, one could argue, is implicit in working with *Anglo-American Cataloguing Rules 2* and *Resource Description and Access*. In one way or another, it seems that a great portion, if not all, of the librarian's work has always been focused on making connections, whether in the construction of a cross reference or, now, in the application of a URI used to interconnect data that:

- differentiates clearly between the conceptual nature of a resource and its individual—*instance of*—physical manifestations;

- unambiguously identifies information entities in the development and use of authorities; and

- uses and exposes the relationships, or connections, that exist between and among entities.

The Linked Data movement, then, is the information community's engagement in using emergent standards and technologies that support the inter-linking of structured data. Not surprisingly, many librarians, aside from being early adopters, are guides to those standards and components that drive toward improving and making seamless how humans and machines interact with the Web to make its resources as accessible as possible.

Engaging in linking data translates to attempting to relate previously unrelated, but not un-relatable, data (anything from a library's operating hours on its home page to the descriptive terms used in a catalog entry used to identify a content item). Linking data means using the set of techniques grounded in standard formats developed for publishing structured data on the Web. Data, when linked, potentially produces knowledge and advances the growth process of the web of data.

Librarians linking data, in effect, enable the World Wide Web to be accessed as if it were a database simultaneously queried from diverse, multiple sources and driven by making connections between datasets. Inherent in linking data is the development of easier and more efficient processes using standards and technologies designed for data management on the Web.

The Semantic Web is, in fact, a movement toward defining a data aggregation process characterized by its immense size and variety and all too obvious to librarians. Whether managing titles, authority files, date formats, alpha-numeric identifiers, and other metadata elements, when data formats are standardized, there is the expectation that resources not only become more readily accessible but, in turn, become more useful and re-useable or re-purposable. Standardization, then, is considered the key to making the "Web of Data" a useful and viable resource, while relating datasets becomes one of its key activities.

3.1 THE W3C LIBRARY LINKED DATA INCUBATOR GROUP

Any coverage of the topic of libraries and Linked Data should begin with the work of the W3C Library Linked Data Incubator Group. Chartered from May 2010 through August 2011, the group's mission was:

> *"To help increase global interoperability of library data on the Web, by bringing together people involved in Semantic Web activities—focusing on Linked Data – in the library community and beyond, building on existing initiatives, and identifying collaboration tracks for the future* (Library Linked Data Incubator Group)."

In the incubator group's 2011 report, Linked Data is described as data that is explicitly conveyed using such standards as RDF and URI. Such standards enable libraries to make assets "more visible and re-usable (Library Linked Data Incubator Group)" on the Web, and beyond their original context, by using Semantic Web and Linked Data standards and principles.

Of particular value to librarians seeking to gain a perspective on the variety of ways in which Linked Data can be of value are the report's overviews of initiatives undertaken by libraries at all collection levels. Establishing a foundation for communicating the benefits of Linked Data, the report covers not only extant initiatives, but, makes future recommendations, and summarizes technologies and resources.

The report includes the following definition of Linked Data echoing Berners-Lee's "rules":

> *"'Linked Data' refers to data published in accordance with principles designed to facilitate linkages among datasets, element sets, and value vocabularies. Linked Data uses Uniform Resource Identifiers (URIs) as globally unique identifiers for any kind of resource, analogously to how identifiers are used for authority control in traditional librarianship…Linked Data is expressed using standards such as Resource Description Framework (RDF), which specifies relationships between things; relationships that can be used for navigating between, or integrating, information from multiple sources* (Library Linked Data Incubator Group)."

Library Linked Data, then, is the published digital information of all types conforming to the principles of linking datasets, element sets, and knowledge-domain vocabularies (vocabularies that provide metadata attribute values) that libraries produce or curate in order to describe resources or resource discovery aids.

Potentially, Linked Data extends the library mark on the Web by making resources available to an increasing end-user base. Extending that mark also means extending the required effort to remove the uncertainty of meaning of natural language for efficient and accurate machine-assisted processing. The need to contextualize multiplying resources using various identifiers potentially leads to the development of open frameworks useful to the library, archive, and museum resource communities.

> *"By using Linked Open Data, libraries will create an open, global pool of shared data that can be used and re-used to describe resources, with a limited amount of redundant effort compared with current cataloging processes* (Library Linked Data Incubator Group)."

Moreover, the work of developers and vendors of library applications, standardized data formats, though not library-centric, still specify data attributes and values used in cataloging resources. One might look at Linked Data's potential as the realization of directing data to the Web in as generally understandable a way as possible. With HTTP and RDF, developers do not need to use domain-specific or proprietary software and may use, or re-use, a growing tool and open-source solution set. Building new services using library data will be, potentially, easier and more efficient, while opening the field of systemic library support to a wide-ranging community of information management and technology professionals.

Linked Data initiatives are characterized by sharing, extending, and re-using data, including that which is multilingual; providing user services functionality; collaboratively and contributively describing resources at the hyper-local and individual levels; and using identifiers that allow a single resource to be described numerous ways.

Globally accessed, unique identifiers—regulated by the Domain Name System, or DNS—designate works, applications, people, places, events, subjects, and other concepts and objects in which one might find interest. By allowing resource citations to span data sources, metadata descriptions become more accessible. Resource provenance and administration, in context then, is made clearer.

What is so unique in unique identifiers? They allow data providers to segment larger documents into discrete, descriptive parts or portions, in other words, unique parts. From these, a single resource can, potentially, be described comprehensively in terms of various languages, identifying numbers, authority data, or any other attribute supporting library services. The result is connected authoritative sources that reduce redundant metadata description.

Linking data is considered a key activity in supporting and improving the end-user experience. As seamlessly Linked Data increases, indexes expand and browsing pathways proliferate improving resource discovery and navigation. With structured data associated to formal vocabularies and definitions, linking data might best be viewed as a bottom up process dependent on communities of collaboration or decentralized contributors, as stated earlier, that produce granular data. The production and aggregation of data may then be visualized using graphs that help end-users visualize the associations between resources.

When libraries link data, it is done with a goal of helping the larger community of libraries, across a range of curation and management processes, in maintaining stronger links between metadata and the entities described by metadata. By linking data, libraries "open up" or "reveal" processes to a receptive and supportive community of conceptualists and developers contributing to the development of solutions to library data management issues. Moreover, because Linked Data is "cloud-friendly" library Linked Data is freed of stand-alone systems and made more accessible.

The Library Linked Data Incubator group gathered and reviewed use cases and case studies submitted by libraries and individuals from around the world. The submissions chosen by the group were organized into three categories of datasets, vocabularies, and metadata element sets. Use cases and case studies that demonstrated "successful implementation of Semantic Web technologies in libraries and related sectors (Library Linked Data Incubator Group)" were chosen for review. Library Semantic Web initiatives falling outside the definition of use cases and case studies, but demonstrating the breadth of potential and beneficial application, were also collected and vetted. Use cases that demonstrated the value of Linked Data technologies (used to describe library resources and contexts) were covered for the purpose of "sharing these descriptions among institutions and with the broader public (Library Linked Data Incubator Group)."

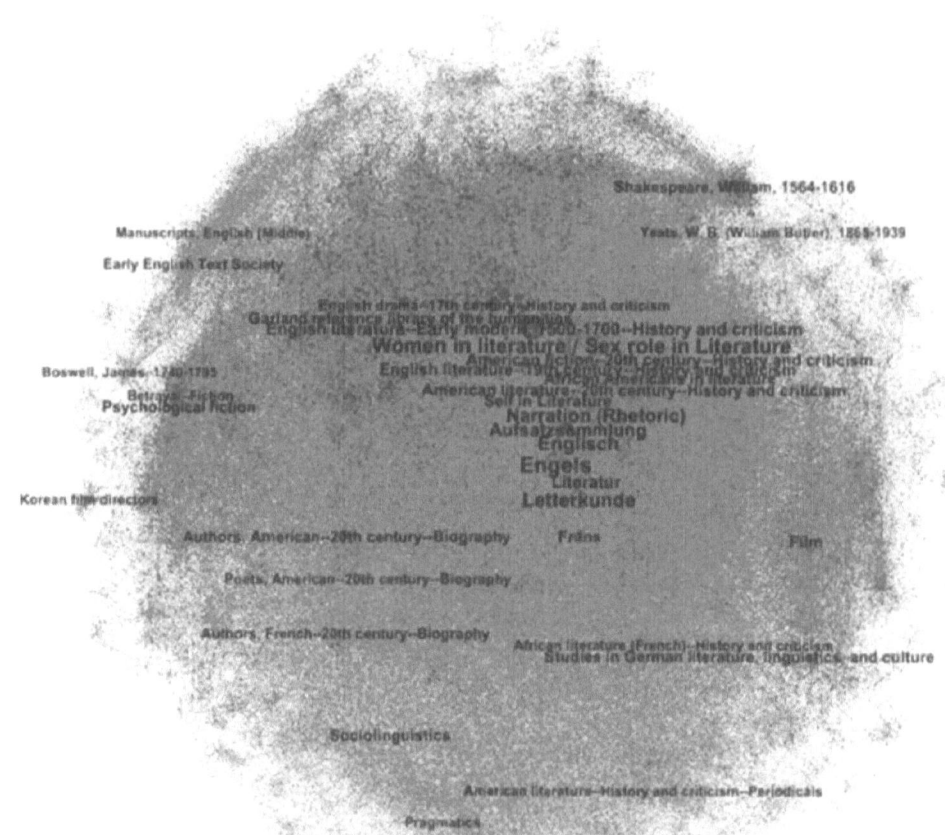

Figure 3.1: Full network graph of all headings in the Literatures and Languages Library at the University of Illinois at Urbana-Champaign (UIUC). Courtesy: UIUC Library (based on: https://www.nics.tennessee.edu/files/images/full-network-graph.jpg).

The report identified three potential Linked Data benefits to libraries:

- improving interoperability of library data by re-using metadata element sets and vocabularies providing descriptive metadata values at a global level;

- applying URIs for resources; and

- developing services, such as APIs, for publishing.

The use cases, case studies, and other initiatives reviewed by the group were organized into the three categories mentioned earlier of datasets; metadata element sets; and, value vocabularies defining "resources (such as instances of topics, art styles, or authors) that are used as values for elements in metadata records (Library Linked Data Incubator Group)."

I. Datasets

Datasets include initiatives such as:

a. The British National Bibliography (BNB) BNB makes accessible the British National Bibliography as linked open data made available through SPARQL services using two different SPARQL interfaces and plain text search functionality.

b. The Open Library

A bibliographic database augmented and maintained to provide "a URI for a work, an edition or author or other book-ish resource that can be used as a pointer and connector for information about books" as well as leading an end-user to a definitive digital version of a work as exists on the Web (Gee and West, 2014)".

c. CrossRef DOI Resolver

CrossRef is a not-for-profit organization founded on publisher collaboration, with a mandate to make reference linking throughout online scholarly literature efficient and reliable and develop other services that are best achieved through collaboration. CrossRef is a DOI Registration Agency and is committed to long-term sustainability. "DOI" refers to Digital Object Identifiers, which are the foundation of a "persistent identifier strategy used by around 3,000 publishers to identify their documents, mostly scholarly publications (CrossRef).

For more information on extant datasets, see http://www.w3.org/2005/Incubator/lld/XGR-lld-vocabdataset-20111025/#Published_Datasets

II. Value vocabularies

Value vocabularies include such examples as:

a. Library of Congress Subject Headings (LCSH)

Obviously, LCSH are well known; however, they might not be quite as well known as a "value vocabulary"—a source of values used as metadata attributes to define the topics of given works—as much as they are considered a comprehensive list of subject headings, published in print and as Linked Data.

b. Getty's Art and Architecture Thesaurus (AAT), a multi-lingual controlled vocabulary, is used to index, catalog, and support search by defining art styles used in the discourse of architecture, the decorative arts, archival materials, and material culture.

c. The Virtual International Authority File (VIAF), a multi-national library joint initiative, brings together name authority files to create what might be described the ultimate name authority service. To date, the service is comprised of more than 20 authority files covering personal, corporate, and conference names.

For more information on extant value vocabularies, see http://www.w3.org/2005/Incubator/lld/XGR-lld-vocabdataset-20111025/#Published_value_vocabularies.

III. Metadata element sets

Metadata element sets include such examples as:

a. Dublin Core and DCMI Metadata Terms

One of the most well-known metadata element sets, DCMI's fifteen "generic property elements" form a base for describing information resources using the namespace http://purl.org/dc/elements/1.1/.

b. Open Archives Initiative—Object Reuse and Exchange (OAI-ORE)

The Open Archives Initiative Object Reuse and Exchange model consists of defined elements describing aggregations of Web resources to form "complex digital objects," for example, journal articles and their digital variations and additional materials.

c. International Standard Bibliographic Description (ISBD)

ISBD, used for describing bibliographic resources of any catalog type, provides a preliminary registration of classes and properties.

For more information on extant metadata element sets, see http://www.w3.org/2005/Incubator/lld/XGR-lld-vocabdataset-20111025/#Metadata_Element_Sets.

3.2   LC LINKED DATA SERVICE

The Library of Congress' Linked Data Service provides its authority data in formats for both human and machine consumption. The Linked Data movement has direct ties to the service with the shared intent of surfacing and inter-connecting data on the Web via "dereferenceable URIs (Library of Congress, "LC Linked Data Service)."

Not surprisingly, the Library of Congress released the LCSH dataset first as part of its service in 2009; and, of the broadest use to librarians, it now makes available the following datasets:

- Library of Congress Subject Headings (LCSH);

- Library of Congress Name authorities;

- Library of Congress Classification; and

- Library of Congress Thesaurus for Graphic Materials.

In terms of formats, the Linked Data service presents data using MADS/ RDF, SKOS, and its proprietary ontology. Various versions are also available such as JSON, N-Triples, RDF/XML/, and XHTML/RDFa (Ford, 2013).

3.2.1   BIBFRAME

Bibliographic Framework Initiative or BIBFRAME, first discussed in Chapter 2, debuted in May 2011 with the intent of furthering the redefinition and launch of a bibliographic data environment developed specifically for the 21st century library landscape.

BIBFRAME is a recent manifestation of more than a half-century of library engagement in electronically sharing cataloging information. BIBFRAME's focus is not on aggregating data about a resource in independently managed bibliographic records; instead, it engages in the development, management, and leverage of relationships existing resource to resource, resource to authority, and so on, and the use of controlled vocabularies.

> "*The BIBFRAME Model is the library community's formal entry point for becoming part of a much larger web of data, where the links between things are paramount* (Library of Congress, 2012)."

This idea of an "entry point" to a vast "web of data" is where libraries begin to join forces with the likes of Google, Bing, Yahoo!, and Yandex all of which work, under the aegis of Schema.org—a schema collection—used by web developers to construct HTML pages findable by the leading search provider and structure data for interoperability (Schema.org).

BIBFRAME advocates that "cultural heritage" organizations, such as libraries, museums, and archives, help end-users find information by actively engaging in assisting search engines and other applications pursue locating information curated by libraries. BIBFRAME, then, has been intentionally designed to harmonize library created cataloging with metadata that libraries also create.

Libraries engaged in the BIBFRAME initiative take advantage of persistent identifiers referencing and contextualizing unique entities in the process of increasing their participation on the larger "Web of Data." Capturing and recording descriptions of library resources is augmented by identifying and establishing additional relations between and among resources. This applies to not only Web resources but to those resources that exist outside "the traditional bounds of the library universe (Library of Congress, 2012)."

Following the tenets of the Semantic Web and linking data, the BIB-FRAME model includes relations or links that drive and transform the web to connect resources across silos, independently maintained, resulting in a graph of related data (see https://www.nics.tennessee.edu/files/images/full-network-graph.jpg) spreading in all directions. Again, search engines and other services use relations between entities to improve the retrieval of relevant information and assist end users in pursuit of that information that will best support their needs.

Relations are essential. One only need look at the Google Knowledge Graph to see the potential of the subject-predicate-object triple when suggesting other resources that may be of interest to an end user. Similarly, the developers of BIBFRAME point out that it is the explicit relationship existing between an individual's buying patterns and Amazon book titles that underpin the calculation of additional item recommendations. Since librarians are experts in determining and establishing bibliographic relationships, the BIBFRAME model seeks to surface their expertise. In summary, BIBFRAME is:

- the starting point of a vision to support numerous content model implementations;

- an enabling system that potentially exchanges data between organizations seamlessly;

- a model that supports emerging metadata rules and content standard; and

- a vehicle for broadening and narrowing the realm of formats used for exchanging bibliographic data (Library of Congress, 2012).

3.3   OCLC LINKED DATA

Library cooperative OCLC is actively engaged in linked data discussions. It is designing and implementing "new approaches that re-envision, expose and share library data as entities that are part of the web." One initiative contributing to OCLC's Linked Data strategy is Linked Open Data (LOD), a project with the goal of extending the Web's data commons by publishing various open data sets as RDF with links between data items from different data sources.

OCLC's Linked Data strategy evolves with periodic exposure of new data and services, while new releases reflect increased understanding of developing information models and publishing information (OCLC, "Data Strategy and Linked Data").

3.3.1   WORLDCAT LINKED DATA

WorldCat's bibliographic metadata is the source of objects that can be represented by linking data. For example, "WorldCat Identities," an initiative which, to date, has created approximately 30 million unique named person, organization, and fictitious character summary pages, are derived from WorldCat data. In WorldCat, bibliographic data are structured according to Schema.org ontologies, or semantic data models, providing WorldCat with the means to offer:

> *"...fullest coverage, improved functionality and better consumption of this data for our harvest partners and library members looking to do more with structured data exposed within HTML pages* (OCLC, "Data Strategy and Linked Data")."

Other OCLC Linked Data initiatives include:

- Dewey.info, a prototype platform and experimental space for linked Dewey Decimal Classification data;

- FAST Linked Data which uses the FAST authorities in linking to corresponding LCSH authorities with many geographic headings linking to the GeoNames geographic database; and

• Virtual International Authorities File (VIAF) an OCLC-hosted name authority service that combines numerous name authority files.

3.4 ADDITIONAL LIBRARY AND VENDOR INITIATIVES

3.4.1 VIVO

The Texas A&M University Libraries launched VIVO in 2014 as a Web-based community of research profiles to enhance faculty collaboration. VIVO provides standard research profiles for all university faculty and graduate students. Researchers can use VIVO to discover and contact individuals with similar interests whether they are across campus or at another VIVO institution. A goal of VIVO is to provide an easy means for researchers and their work on the Web, to be identified. This identification then allows researchers to locate and connect potential collaborators in the development of interdisciplinary projects.

VIVO's implementation complements the Libraries' roll out of Open Researcher and Contributor ID (ORCID), a registry of unique identifiers for researchers, that is part of an expanding suite of services offered by the Texas A&M libraries to develop research opportunities and scholar citations through focused applications and software. VIVO is also intended to support the development of collections and services that support researchers' needs by recognizing scholarly expertise across campus (VIVO).

3.4.2 POLYMATH VIRTUAL LIBRARY

The aim of Polymath Virtual Library is to unite information, data, digital texts, and websites covering the philosophy, politics, and science of Spanish, Hispano American, Brazilian and Portuguese polymaths across time and in all languages (W3C, 2011).

3.4.3   TALIS PRISM

Powered by the Talis platform consisting of a hosted Linked Data service offering SPARQL and full text search, is "a next-generation OPAC/search and discovery interface." Recognizing that "there is a need to offer a rich interface to surface the large volume of content available in libraries," Talis Prism supports browse of author, subject, and series element values (W3C, 2010).

3.4.4   SFX

Moving outside the library to the domain of vendors, SFX—a service of library automation provider, Ex Libris—mediates the relations between resources using "link resolvers" or "link-servers," which divide or separate the elements of an OpenURL (i.e., a URL in a standardized format that allows end-users to find resources that they are allowed to access) while providing links to "targets" in libraries using an OpenURL knowledge base of electronic resource information indicating the availability and accessibility of electronic resources, e.g., electronic journals or ebooks (Ex Libris).

3.5   SUMMARY

In this chapter, we examined some of the Semantic Web initiatives in which libraries and related enterprises are engaged. With a particular focus on Linked Data, we provided a high level overview of library or vendor initiatives using technologies covered in Chapter 2. These initiatives cover:

- providing opportunities to focus library work and resources on ever more specific or local domains;

- aggregating and sharing Web-based identifiers to provide resource descriptions that cross domains of culture and knowledge;

- citing resources using Web-based identifiers; and

- maintaining channels to reusable and separately managed syntactic and semantic data.

CHAPTER 4

Conclusion

Stating the obvious, the Web is an evolving space; and, as a related concept, the Semantic Web evolves. Because no single agency or organization controls all aspects of the Semantic Web's development and use, spirited debates covering the direction in which the Semantic Web is "evolving;" what "form" it will "take;" or, by which synonym or near-synonym it might one day be known are occurring at conferences and in graduate schools, and, most importantly to this lecture, in libraries everywhere, now. This means that libraries have been able to take a certain lead in contributing to and forming the guidelines and protocols that comprise the Semantic Web.

Still, much of the Semantic Web's function and practicality are still in development with challenges to overcome. Because the Semantic Web is decentralized, its developers have the freedom to create the descriptors and vocabularies that they need, instead of adhering to the Semantic Web approach of reusing data structures, semantic relations, and vocabulary values. Also, some critics of the Semantic Web consider its vision to be impractical and question the likelihood that enterprises, using available tools for managing metadata, will devote significant resources to adding all the necessary metadata to content. But, as we have seen, libraries have been addressing such challenges for a long time:

> *"The library community has the right intentions about creating a new data interchange format/standard to help them share with the wider web bibliographic data…So the ambitions are right but probably the average person not in the library community won't get BIBRAME when it's finished* (Zaino, 2013).*"*

Hopefully, what the "average person" will "get" or appreciate is, among other innovations, a better end-user experience particularly as it relates to search. To get there, however, the library community must continue to explore the Semantic Web's possibilities and ask the questions needed to make it viable.

Bibliography

Antoniou, G. and F. van Harmelen. "Preface." A Semantic Web Primer, edited by John Mylopoulos and Michael Papzoglou. Cambridge, US: The MIT Press, 2004. http://www.google.com/url?sa=t&rct=j&q=&esrc=s&source=web&cd=2&ved=0CC4QFjAB&url=http%3A%2F%2Fwww.dcc.fc.up.pt%2F~zp%2Faulas%2F1011%2Fpde%2Fgeral%2Fbibliografia%2FMIT.Press.A.Semantic.Web.Primer.eBook-TLFeBOOK.pdf&ei=7-1pVKrCN5XasATY7YLIAQ&usg=AFQjCNFpAkxja-J5hzNoMxb8kQAif_EtDcQ&bvm=bv.79142246,d.cWc. 1

Bergsell, B. "The difference between the Web and the Semantic Web." Swipnet, 08 Aug. 2001. http://home.swipnet.se/semanticweb/thesis/parts/sw_diff_web.html. 1

Berners-Lee, T. "Linked Data." *W3C*, 18 June 2009. http://www.w3.org/DesignIssues/LinkedData.html. 23, 25

Berners-Lee, T., J. Hendler, and O. Lassila. "The Semantic Web: A new form of Web content that is meaningful to computers will unleash a revolution of new possibilities." *Scientific American*, 17 May 2001. 2, 25

CrossRef. "History/Mission." 24 Mar. 2011. http://www.crossref.org/01company/02history.html. 31

Eisenberg, V. "On the difference between Linked Data and Semantic Web." Vadim on Software and Semantic Web, 29 Oct. 2011. http://vadimeisenberg.blogspot.com/2011/10/on-difference-between-linked-data-and.html. 2

Ex Libris. "SFX , the OpenURL link resolver and much more." http://www.exlibrisgroup.com/category/SFXOverview. 38

Feigenbaum, L. and E. Prud'hommeaux. "SPARQL by Example." Cambridge
    Semantics. http://www.cambridgesemantics.com/semantic-university/
    sparql-by-example. 19, 20

Ford, K. "Library of Congress Classification as linked data". JLIS.it. Vol. 4, n. 1.
    (Gennaio/January 2013): Art: #5465. DOI: 10.4403/jlis.it-5465. 34

Gee, D. and J. West. "About Open Library." Open Library, 12 Aug. 2014. https://
    openlibrary.org/about. 31

IETF (Internet Engineering Task Force). Uniform Resource Identifiers (URI):
    Generic Syntax, STD 66, ed. T. Berners-Lee, R. Fielding, and L. Masin-
    ter, 2005. https://www.ietf.org/rfc/rfc2396.txt. 6

Khan, M. "Semantic Web." CSTags, 8 May 2014. http://cstags.com/info-tech/
    semantic-web/. 3

LATC. "The Linking Open Data cloud diagram," maintained by Richard Cyga-
    niak and Anja Jentzsch. 30 Aug. 2014. http://lod-cloud.net/. 21

Library of Congress. "Bibliographic Framework as a Web of Data: Linked Data
    Model and Supporting Services." 21 Nov. 2012. http://www.loc.gov/
    bibframe/pdf/marcld-report-11-21-2012.pdf. 13, 18, 34, 35

Library of Congress. "Linked Data Service: Authorities and Vocabularies." http://
    id.loc.gov/about/. 33

OCLC. "Data Sets & Services." http://www.oclc.org/data/data-sets-services.
    en.html. 18

OCLC. "Data Strategy and Linked Data: Helping libraries thrive on the web."
    http://www.oclc.org/data.en.html. 36

Schema.org, http://schema.org. 34

SemanticWeb.org. "Main Page." 7 Nov. 2012. http://semanticweb.org/wiki/
    Main_Page. 2

VIVO. "What is VIVO?" http://www.vivoweb.org/about. 37

W3C. "Library Linked Data Incubator Group Final Report." http://www.w3.org/2005/Incubator/lld/XGR-lld/. 12, 23, 26, 27, 28, 29, 31

W3C. "Linked Data." http://www.w3.org/standards/semanticweb/data. 2, 20, 21

W3C. "OWL Web Ontology Language: Overview." 10 Feb. 2004. http://www.w3.org/TR/owl-features/. 15

W3C. "RDF Schema 1.1.: Recommendation" 2014. http://www.w3.org/TR/rdf-schema/. 12

W3C. "Resource Description Framework (RDF)." http://www.w3.org/RDF/.

W3C. "Use Case Polymath Virtual Library." 17 Jan. 2011. http://www.w3.org/2005/Incubator/lld/wiki/Use_Case_Polymath_Virtual_Library. 37

W3C. "Use Case Talis Prism 3." 16 Nov. 2010. http://www.w3.org/2005/Incubator/lld/wiki/Use_Case_Talis_Prism_3. 38

W3C. "Vocabularies." http://www.w3.org/standards/semanticweb/ontology.html. 17

W3C. "Semantic Web." http://www.w3.org/standards/semanticweb/. 3

Wikipedia contributors. "Linked data." Wikipedia, The Free Encyclopedia, 1 Oct. 2014. http://en.wikipedia.org/wiki/Linked_data. xiii

Wilson, T. "How Semantic Web Works." How Stuff Works, 09 Feb. 2006. http://computer.howstuffworks.com/semantic-web.htm. 3

Zaino, Jennifer. "The Future of Libraries, Linked Data and Schema.Org Extensions." Semanticweb.com, 13 Feb. 2013. http://semanticweb.com/the-future-of-libraries-linked-data-and-schema-org-extensions_b35315. 39

Author Biographies

Keith P. DeWeese has over 20 years of experience as a librarian and practitioner in the areas of taxonomy and ontology development, text analytics, knowledge management, content management, and information retrieval. He has designed and delivered semantic solutions for a variety of industries including news media, healthcare education, and consumer products, as well as for the government and nonprofit sectors. Keith holds an M.L.S. degree from Florida State University.

Dan Segal is a Senior Taxonomist with Marcinko Enterprises, Inc. Dan has over 20 years of experience as a practitioner and consultant in the areas of taxonomy and ontology development, text analytics, knowledge management, content management, and information retrieval. He has designed and delivered semantic solutions for a variety of industries including healthcare, pharmaceutical, consumer products, and media, as well as for the government and nonprofit sectors. Prior to joining MEI, Dan was Manager of Taxonomy Delivery at Dow Jones Client Solutions and Associate Director of Information and Knowledge Integration at Bristol-Myers Squibb. He holds B.S. and M.L.S. degrees from Rutgers University.